Holt Social Studies

World Geography

HOLT, RINEHART AND WINSTON

A Harcourt Education Company

Orlando • **Austin** • New York • San Diego • Toronto • London

ISBN 0-03-078042-X

1 2 3 4 5 6 7 082 08 07 06 05

Contents

Northern Europe

Vocabulary Builder

Section 1

Arctic Circle	British Isles	fjord
geothermal energy	glaciers	hydroelectric energy
North Atlantic Drift	peninsula	Scandinavia
subarctic	tundra	volcanoes

DIRECTIONS Read each sentence and fill in the blank with the word in the word pair that best completes each sentence.

1. Many lakes in the British Isles were carved by ___________________ millions of years ago. (**volcanoes/glaciers**)

2. Northern Europe is made up of the British Isles and ___________________. (**Scandinavia/the Artic Circle**)

3. A ___________________ is a narrow inlet of the sea set between high, rocky cliffs. (**fjord/peninsula**)

4. Energy produced from the heat of Earth's interior is called

   ___________________. (**geothermal energy/hydroelectric energy**)

5. Northern Europe experiences a mild climate due to the

   ___________________. (**Arctic Circle/North Atlantic Drift**)

DIRECTIONS Look at each set of four vocabulary terms. On the line provided, write the letter of the term that does not relate to the others.

______ **6. a.** glacier
      b. rainfall
      c. fjord
      d. cliff

______ **8. a.** Ireland
      b. England
      c. Scotland
      d. Denmark

______ **7. a.** tundra
      b. soil
      c. forest
      d. ocean

______ **9. a.** North Atlantic Drift
      b. mild climate
      c. ocean current
      d. subarctic climate

Northern Europe

Vocabulary Builder

Section 2

agreement	constitutional monarchy	disarm
Dublin	Industrial Revolution	London
Magna Carta	Stonehenge	

DIRECTIONS Read each sentence and fill in the blank with the word in the word pair that best completes the sentence.

1. The _________________________ was characterized by the growth of industries like textiles, iron, and steel. (**Industrial Revolution/Magna Carta**)

2. An _________________________ led to the creation of a national assembly and a cease fire in Northern Ireland in the late 1990s. (**agreement/Industrial Revolution**)

3. _________________________ is an ancient monument in the British Isles that was built 5,000 years ago. (**Dublin/Stonehenge**)

4. A _________________________ is a government that has a monarch and a legislative body that makes the laws. (**Magna Carta/constitutional monarchy**)

5. Peace talks in Northern Ireland were stalled when some groups refused to

_________________________. (**disarm/agreement**)

DIRECTIONS Write three words or phrases that describe the term.

6. London ____

7. Dublin ____

8. Magna Carta ____

Northern Europe

Vocabulary Builder

Section 3

domination	geysers	Helsinki
hot springs	neutral	Oslo
Stockholm	uninhabitable	Vikings

DIRECTIONS Answer each question by writing a sentence that contains at least one word from the word bank.

1. How were longboats used by Scandinavian warriors?

____

____

2. Why do most people in Greenland live on the southwest coast?

____

____

3. What role did Sweden play in World War I and World War II?

____

____

4. How are many homes in Iceland heated?

____

____

5. What city in Scandinavia is often called a floating city and why?

____

____

Northern Europe Biography

Mary, Queen of Scots
1542–1587

HOW SHE AFFECTED THE WORLD Mary, Queen of Scots is Scotland's most famous queen, known for her beauty and charm. She was a successful ruler for a time. Eventually, she was forced out of Scotland and beheaded in England at the age of 44.

As you read the biography below, think about how likeable Mary must have been to win over the Scottish people.

Mary Stuart was only six days old when her father died, making her Queen of Scotland. At age five, Mary's mother, a Frenchwoman, sent her to France to be brought up at the court of King Henry II. At that time, she was **betrothed** to the heir to the French throne.

During her time in France, Mary became a true **Renaissance** princess, educated in several languages, music, and poetry. She also loved outdoor activities such as golf, archery, and hunting. Through her marriage to Francis II in 1558, Mary actually became the queen of France. It was short-lived, however, as Francis died in 1560.

Mary returned to Scotland the next year, only to find that she was a Catholic queen in a Protestant country. Because of her French education and manners, Scottish people thought of her as a Frenchwoman. Despite these challenges, she won the people over and ruled successfully for a time.

Mary's downfall started with her marriage to her English cousin, Lord Darnley, in 1565. Lord Darnley was unpopular with the people. Royal counselors advised Mary to get rid of him. In 1567 Darnley

VOCABULARY

betrothed arranged to marry

Renaissance period of cultural, scientific, and artistic change in European history

treason act of betraying one's country

was murdered. A few months later, Mary married the prime suspect. This made her unpopular with the Scots, and in 1567, she was imprisoned and forced to give up the throne to her infant son, James. In 1568, she fled to England.

Queen Elizabeth I, a cousin of Mary's, kept Mary imprisoned in England rather than turning her over to the Scottish government. Over the next 18 years, Mary tried unsuccessfully to escape. Mary was also linked to many plots to overthrow or kill Queen Elizabeth, as her family lineage gave her a claim to the English throne. Tired of the constant threat, Elizabeth eventually had her cousin tried for **treason.** Mary was beheaded in 1587.

WHAT DID YOU LEARN?

1. Analyze and Make Judgments How did Mary's choice in husbands affect her life?

____

____

2. Drawing Conclusions Why do you think Mary may have plotted against Queen Elizabeth?

____

____

ACTIVITY

Conduct additional research about Mary Stuart's life, and create a detailed family tree to show how she was related to other rulers of the time.

Northern Europe   Biography

Hans Christian Andersen
1805–1875

HOW HE AFFECTED THE WORLD Hans Christian Andersen is Denmark's most famous author. His fairy tales have been read and loved by adults and children around the world since his first collection appeared in 1835. Many of his stories, which include "The Ugly Duckling" and "The Princess and the Pea," are autobiographical.

As you read the biography below, think about the perseverance that helped Andersen become so successful and famous.

VOCABULARY

autobiographical written by a writer about him- or herself

mad insane

eccentric someone thought to be odd or strange

eloquence ability to express oneself well

Thousand and One Nights a very old collection of Arabic stories (also called *Arabian Nights*)

Hans Christian Andersen's own life was much like one of his fairy tales. He was born in Odense, Denmark, the son of a poor shoemaker who died when Andersen was eleven. His grandfather was known as a **mad eccentric**. His grandmother worked in the local insane asylum and occasionally allowed Andersen to go to work with her. Near the asylum, the poor townswomen spun thread. Andersen explained, "I often went in there and was very soon a favorite . . . I passed for a remarkably wise child, that would not live long; and they rewarded my **eloquence** by telling me tales in return; and thus a world as rich as that of the ***Thousand and One Nights*** was revealed to me."

Andersen went to Copenhagen at the age of 14 in hopes of becoming a famous actor. Although that plan failed, he met Jonas Collin, who was associated with the Royal Theatre. Collin took Anderson under his wing and secured a royal grant for him to continue his education.

Hans Christian Andersen, *continued*   Biography

He was admitted to the University of Copenhagen in 1828. The next year, he published his first important literary work, *A Journey on Foot from Holmen's Canal to the Eastern Point of Amager.* This fantastic tale was an instant success.

Andersen then turned to playwriting and poetry, but was mostly known as a novelist. In 1835, he published his first book of fairy tales, *Eventyr.* (In Dutch, an *eventyr* is a kind of fantastic tale, not solely intended for children.)

While some of Andersen's tales end happily, many are tragic and pessimistic. He wrote 156 fairy tales in all. Some of his most widely recognized stories include "The Ugly Duckling," "The Princess and the Pea," "The Snow Queen," and "The Emperor's New Clothes."

WHAT DID YOU LEARN?

1. Analyze and Make Judgments How did Hans Christian Andersen's move to Copenhagen affect his life?

____

____

2. Using Deductive Reasoning How was Hans Christian Andersen's own life similar to a fairy tale?

____

____

ACTIVITY

Conduct additional research on the life of Hans Christian Andersen. Then find one of his fairy tales that is autobiographical. Pretend you are a Danish newspaper reporter interviewing Andersen about the connection between his life and this fairy tale. Write your interview on a separate piece of paper.

Romeo and Juliet
by William Shakespeare

ABOUT THE READING William Shakespeare is often considered the most important playwright in the history of British literature. He was also an actor and poet. Shakespeare's plays are divided into comedies, histories, and tragedies. *Romeo and Juliet* is one of his most famous tragedies. His plays and **sonnets** are written in **iambic pentameter**.

VOCABULARY

sonnet a poem with fourteen lines

iambic pentameter rhythmic pattern that alternates between unstressed and stressed syllables

doff shed, give up

counsel private thoughts

As you read the passage below try to pay attention to the rhythm of each line.

Juliet: [*not knowing Romeo hears her*] O Romeo,
Romeo, wherefore art thou Romeo?
Deny thy father and refuse thy name,
Or if thou wilt not, be but sworn my love,
And I'll no longer be a Capulet.
Romeo: [*aside*] Shall I hear more, or shall I speak at this?
Juliet: 'Tis but thy name that is my enemy.
Thou art thyself, though not a Montague.
What's Montague? It is nor hand, nor foot,
Nor arm, nor face, nor any other part
Belonging to a man. O, be some other name!
What's in a name? That which we call a rose
By any other word would smell as sweet.
So Romeo would, were he not Romeo called,
Retain that dear perfection which he owes
Without that title. Romeo, **doff** thy name,
And for thy name—which is no part of thee—
Take all myself.
Romeo: [*to Juliet*] I take thee at thy word.
Call me but love and I'll be new baptized.
Henceforth I never will be Romeo.

> What is Romeo's last name? What is Juliet's?

> What does Juliet's reference to a rose mean?

> What does Romeo mean by the use of the word *baptized*?

Source: *The Norton Shakespeare*, edited by Stephen Greenblatt. New York: W.W. Norton & Company, 1997

Juliet: What man art thou that, thus bescreened in
    night,
    So stumblest on my **counsel**?
Romeo: By a name
    I know not how to tell thee who I am.
    My name, dear saint, is hateful to myself
    Because it is an enemy to thee.
    Had I it written, I would tear the word.
Juliet: My ears have yet not drunk a hundred words
    Of thy tongue's uttering, yet I know the sound.
    Art thou not Romeo, and a Montague?
Romeo: Neither, fair maid, if either thee dislike.
Juliet: How cam'st thou hither, tell me, and
    wherefore?
    The orchard walls are high and hard to climb,
    And the place death, considering who thou art,
    If any of my kinsmen find thee here.

ANALYZING LITERATURE

1. Main Idea What is the problem that Romeo and Juliet are discussing?

____

____

2. Critical Thinking: Drawing Inferences How do you think Romeo and Juliet will
proceed with their relationship? Why?

____

____

ACTIVITY

Conduct research on both iambic pentameter and sonnets in order to
fully understand each term. Now write your own sonnet in iambic
pentameter about some aspect of Northern Europe.

Northern Europe

"The Scream" by Edvard Munch

ABOUT THE ARTIST Edvard Munch was born in Oslo, Norway, in 1863. His work, an **antecedent** of the **expressionist** movement, often depicted shocking themes of fear and death. "The Scream" is one of his most famous works.

VOCABULARY

antecedent something that comes before

expressionist representing feelings and moods rather than reality

temperament the way a person thinks, behaves, and reacts

As you examine the painting, consider the feelings and emotions the artist is trying to convey. Notice how the painting makes you feel.

"The Scream," *continued*   Primary Source

WHAT DID YOU LEARN?

1. What physical feature common to the Norwegian coast is depicted in the painting?

____

____

2. Use at least three adjectives to describe how the person is feeling in the painting and give reasons for your answer.

____

____

3. How would you describe the person's appearance?

____

____

4. How is nature depicted in this painting?

____

____

Northern Europe   Geography and History

Understanding Northern Europe

The geography of Northern Europe makes it a place of extreme physical diversity. You can find rolling hills, high cliffs, volcanoes, and shooting geysers. You might see lush valleys, rocky terrain, or areas where the climate is so cold that only grass and moss can grow.

MAP ACTIVITY

1. Color the countries of the United Kingdom with a yellow pencil.

2. Use a green pencil to shade the countries of Scandinavia.

3. Use a red pencil to outline at least three fjords.

4. Place an "X" in all areas on the map that were once conquered or settled by the Vikings.

ANALYZING MAPS

1. **Movement** About how many miles would the Vikings have sailed to get from Sogne Fjord to the Gulf of Bothnia?

____

____

Understanding Northern Europe, *continued* Geography and History

2. Location How do you think the location of Northern Ireland contributes to the conflict there?

____

____

3. Location Describe the location of Northern Europe in terms of latitude and longitude.

____

____

4. Comparing and Contrasting Why do you think the east coast of the Scandinavian Peninsula is so different in appearance from the west coast?

____

____

Northern Europe   Social Studies Skills

Study

Writing to Learn

LEARN THE SKILL

Writing is an active process that can help you learn new information.
The act of writing helps focus your attention. It can also encourage
you to think analytically about a particular topic. Putting something in
your own words helps you discover what you do and do not understand
about a subject.

Some students may prefer to jot down quick notes or make outlines.
Others may learn better by writing more extensively, in paragraph-style.
Usually, writing to learn involves informal writing, such as note-taking,
journal writing, letters, free-writing, or personal responses.

PRACTICE THE SKILL

Read back over your notes from this chapter. Then set a timer for
15 minutes. On a separate piece of paper, write down all the details you
remember from the chapter. Try to write continuously for the full 15
minutes, even if you have to repeat something you already wrote.

If you had trouble writing for the full time, look at your notes from
the chapter once more. Are they so brief that they did not make sense
to you when you reviewed them? Or are they so long that all your focus
was on writing rather than on understanding the information?

The Vikings Abroad

Vikings is the term we use today for the people who, from the AD 700s to the 1000s, left their homelands in Scandinavia for other lands. They began three centuries of raiding, trading, exploring, and colonizing. The Viking movements are a good example of migration.

REASONS FOR MIGRATION

Many people have wondered what prompted the Viking expansion. People migrate, or permanently relocate, in response to push and pull factors. Push factors are things about a person's current location that are undesirable. These factors may "push" the person to leave. Pull factors are the things about a new location that are attractive and "pull" a person toward it. Not enough land for farming, herding, or hunting was an important push factor for the Vikings. Also, some Vikings wanted to escape the rule of a particular king or chief.

Many historians now think that the increasing wealth of Europe was the biggest pull factor. By the 700s, trade was picking up again after having collapsed with the fall of the Roman Empire. Treasure was piling up in churches, monasteries, and other religious centers. Increased opportunities to plunder attracted the Vikings.

BARRIERS TO MIGRATION

For any group or individual, there are also barriers to migration. These can be physical, cultural, economic, or legal. The major barrier to migration for the Vikings was that the native peoples did not want them in their lands. In some places, the Vikings were defeated militarily. In other places, they raided successfully. They gained treasure and slaves, but no land. In yet other places, the Vikings were able to establish towns and villages and bring in hundreds or thousands of settlers. Some people paid the Vikings to go away. These bribes were called Danegelds. Others offered the Vikings land to settle on; this is what happened in Normandy. Yet other groups invited the Vikings to become their rulers. These groups included the Slavic people in the area of present-day Novgorod (see map). In this way, the Vikings, who were sometimes called the Rus, became the founders of Russia. It is clear from the Viking example that immigrants may receive a variety of responses from others when they arrive.

EFFECTS OF MIGRATION

Migration can have many consequences. For example, the Vikings who settled in the lands where they were raiders or traders mixed into the local society. They learned the local language, married local people, and adopted new religions such as Christianity.

The Viking expansion had many effects on people's lives. Sometimes trade was disrupted, but at other times it was improved. Boat design improved with Viking technology. Numerous place-names in eastern England developed from Viking words. Words such as *window, husband, sky, anger, low, ugly, wrong, happy, thrive, ill, die, bread,* and *eggs* came into English this way.

The Vikings brought democratic ideas with them to Iceland. There they established the Althing, the world's first parliament. Upon arrival, they also immediately set to work transforming the landscape of the island, heavily exploiting the walrus (which apparently died out rather quickly) and the great auk (which eventually became extinct in the 1800s). Wood-cutting and the introduction of domesticated animals and plants led to the destruction of much of the natural vegetation of the island.

The effects of the Viking emigration on their homelands were several. Wealth from raiding and trading poured into the region. Some wealth was undoubtedly carried back to Scandinavia by returning migrants. Some Vikings went back to their old homes, rather than staying in the new land. For every migration stream, there is a return migration stream. People who returned were either disappointed with the new place or they had always intended to go home.

YOU ARE THE GEOGRAPHER

Use the map on the next page and an atlas to answer the following questions.

1. What three modern countries were the Viking homelands?

____

2. From which islands did Vikings go west to Iceland and beyond?

____

3. From which countries did Vikings go south to western and southern Europe?

____

4. From which country did Vikings go east to eastern Europe, Russia, and Asia?

____

5. What is the latest date on the map for an area "occupied or dominated" by Vikings? Where is it? This place is part of what country today?

____

____

Write an essay that compares and contrasts a later example of migration to the Viking example. This could be the migration story of an individual or family with whom you are familiar or of a large group such as Scandinavians coming to the United States after the Civil War or Mexicans drawn to California or Texas in the 1900s. Discuss push and pull factors, barriers to migration, and the consequences of the migration. Explain how your example of migration is similar to and different from the experience of the Vikings.

The Viking Migration

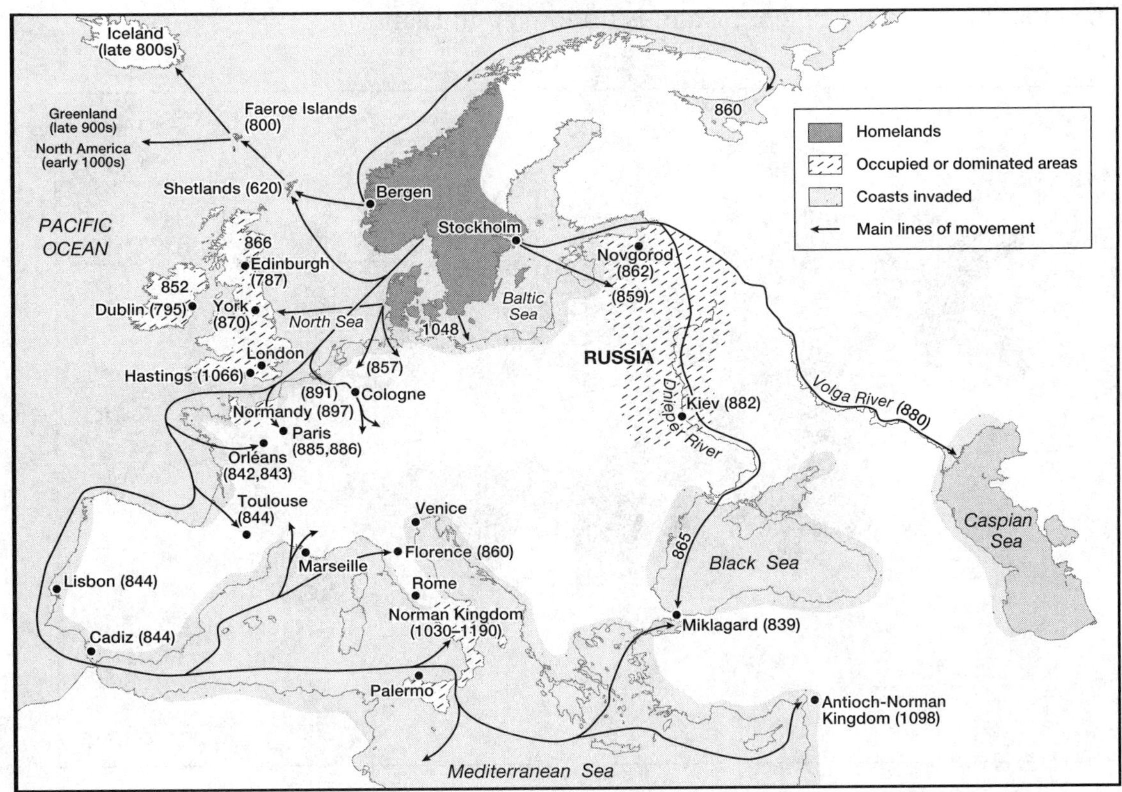

Connecting Europe by Chunnel

In May 1994 Great Britain's Queen Elizabeth and French president François Mitterrand opened the tunnel that connects their two nations under the English Channel. The Channel Tunnel, or "Chunnel" as it is popularly known, is the result of 200 years of hopes and plans to link Great Britain and France. To learn about the Chunnel, study the passage, map, and diagram and answer the questions.

The 31-mile-long Chunnel is really three parallel tunnels: two for trains and a service tunnel. It snakes from Folkestone, England, to Coquelles, France, an average of 150 feet below the seabed. Drive onto a train at one end; stay in your car and drive off . . . at the other [end] 35 minutes later. . . Passenger trains . . . provide through service: London to Paris in three hours; London to Brussels in three hours, ten minutes. The Chunnel rewrites geography, at least in the English psyche [mind]. The moat has been breached. Britain is no longer an island.

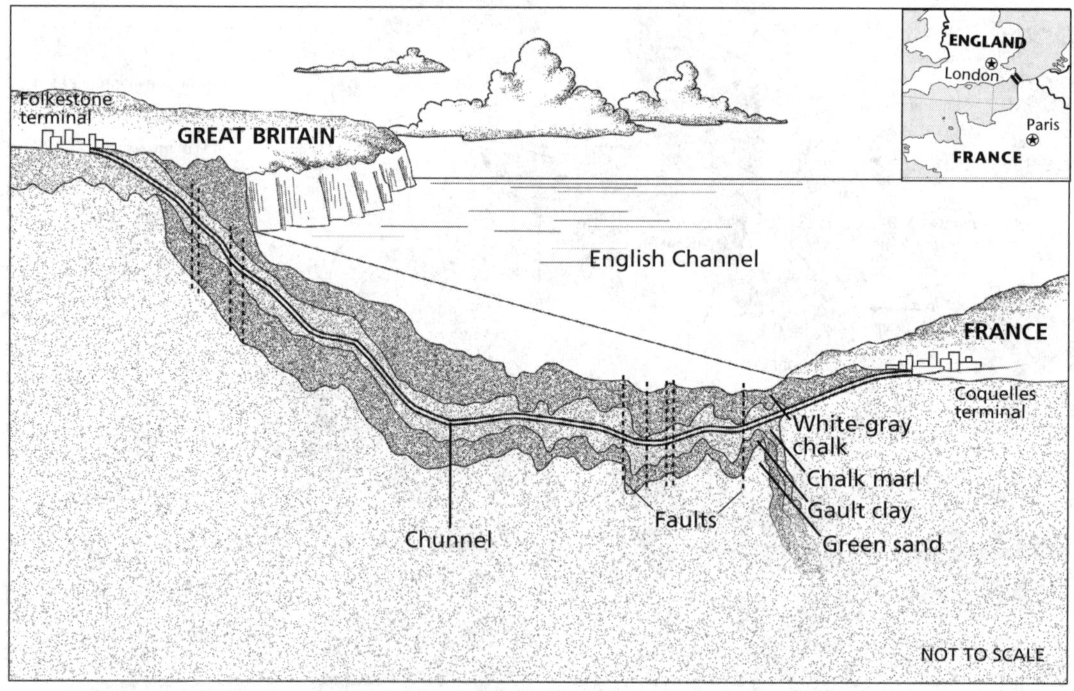

On a clear day . . . on the north coast of France, you can look across the Channel and see the faint shimmer of chalk cliffs on the English coast. It is 22 miles to Dover as the bluebird flies. Ninety minutes by ferry, 40 minutes by hovercraft. . . Odds of a gale on any day in winter are one in seven. Now neither freight nor passengers need be held hostage by a heaving sea.

. . . Dreams of a cross-Channel connection have materialized, only to evaporate, 26 times before. Albert Matheiu, a French engineer, proposed the first plan in 1802. . . Napoleon, Britain's archenemy, approved, and there's the curse. . . Why, after two centuries, did the two ancient foes finally take the plunge? A congruence [alignment] of stars and planets[?] . . . Sir Alastair Morton, the British chairman of Eurotunnel—the Anglo-French company that owns and operates the tunnel [says it is] "more likely a congruence of political tides and currents" . . . On his wall is a copy of the legendary London newspaper headline: "Fog in Channel. Continent Cut Off" . . .

Other converging currents: a tidal wave of 220 banks from 26 nations to finance the project. British and French industries desperate for work (the tunnel [construction] employed as many as 15,000 workers).

"The tunnel's come about because Britain finally feels secure," adds Sir Christopher Mallaby, Britain's ambassador to France. Still, the country just can't go it alone. Britain, which snubbed the rest of Europe for years, has joined the European Union . . . The island mentality has eased. . .

From "The Light at the End of the Chunnel" (retitled "Connecting Europe by Chunnel") by Cathy Newman from National Geographic, May 1994. Copyright © 1994 by National Geographic Society. Reprinted by permission of the publisher.

1. What does the phrase "Britain is no longer an island" mean?

____

2. What does the London newspaper headline suggest about Britons' traditional attitude toward Europe? What impact would this view likely have had on plans for a tunnel?

____

____

3. Why would the engineers have chosen to construct the tunnel in the marl layer rather than choosing a depth that was higher or lower?

____

____

4. What geological dangers to the Chunnel are shown by the drawing? What risks do they present for the future of Chunnel operations?

____

____

Northern Europe   Focus on Reading

Using Context Clues—Synonyms

As you read, you may occasionally encounter a word or phrase that you
do not know. When that happens, use the words and sentences around
the unfamiliar word—context clues—to help you determine the word's
meaning. Sometimes the clues you find will be words or phrases that
have the same meaning as the unfamiliar word. These are synonyms.

PRACTICING USING CONTEXT CLUES
Use context clues to determine the meaning of the underlined word
in the following excerpts from your chapter. Write the meaning of the
underlined word in the space provided.

1. As the glaciers flowed slowly downhill, they carved long, winding channels, or
fjords, into Norway's coastline.

____

2. Iceland also has an unusual resource, geothermal energy, energy produced from
the heat of Earth's interior.

____

3. Early settlers built Stonehenge, an ancient monument, some 5,000 years ago.

____

4. One similarity is their common ancestry. Many people in the British Isles trace
their heritage to the early invaders of the region, such as the Celts, Angles, and
Saxons.

____

5. Oslo, the capital of Norway, is the country's leading seaport, as well as its
industrial and cultural center.

____

____

6. Because some 80 percent of the island is covered in a thick ice sheet, much land
on the island is uninhabitable, or not able to support human settlement.

____

CREATING YOUR OWN CONTEXT CLUES

colonies	fertile	marine
popular culture	populous	primary resources
prosperous	Vikings	

The word bank contains words and terms from your chapter. Choose
five words and use each of them in a sentence about Northern Europe
that includes a context clue.

1. ____

____

____

2. ____

____

____

3. ____

____

____

4. ____

____

____

5. ____

____

____

Writing a Letter

Letters are a great way to stay in touch with family and friends. Gather information about Northern Europe as you read the chapter. Now, imagine you are traveling through this region. Write a letter to your friends and family at home in which you describe what you have seen and learned in your travels.

PREWRITING

1. **Describing the Physical Geography** Take notes on the physical features, resources, and climates of Northern Europe. Include information about the following: fjords, coastlines, terrain, countries of the British Isles, countries of Scandinavia, geothermal energy, primary resources, and the North Atlantic Drift.

2. **Writing about the British Isles** What information about the British Isles could you include in a letter to someone who has never visited the area? Consider the following: early history, the British Empire, the Industrial Revolution, Ireland's famine, popular culture and traditions, disputes, and current economic and political situations.

3. **Writing about Scandinavia** Add information about Scandinavia to your notes. Include details about the following: early history, the Vikings, cultural and economic traits of each country, natural resources, and geysers.

WRITING

4. **Preparing to Write your Letter** Decide how to organize it. You may imagine that you are writing a bit at the end of each day of travel. Or you may imagine that you are coming home on the plane and writing a summary of the trip.

5. **Writing the Letter** Try to let it read like a real letter rather than a report on Northern Europe. Include your imagined reaction to what you have seen and done in Northern Europe. What were your favorite things about your imaginary trip? Was there anything you did not enjoy?

Writing a Letter, *continued*   Focus on Writing

EVALUATING AND PROOFREADING

6. Evaluating your Letter Would someone reading your letter get a good sense of
the geography, history, and personality of Northern Europe, even if he or she
knew nothing about the region?

Rubric Use this rubric to check that you have included all the
required information in your letter.

- Have you included information about the physical geography and climate of
  Northern Europe?

- Have you included details about the history, people, culture, and current issues
  of the British Isles?

- Have you included details about the history, people, culture, and current issues
  of Scandinavia?

- Have you included your personal reaction to the people and places you've
  seen?

7. Proofreading your Letter Last, check the following:

- Capitalization and spelling of all proper names and places
- Punctuation, grammar, and spelling

Northern Europe   Chapter Review

BIG IDEAS

1. Northern Europe is a region of unique physical features, rich resources, and diverse climates.

2. Close cultural and historical ties link the people of the British Isles today.

3. Scandinavia has developed into one of the most stable and prosperous regions in Europe.

REVIEWING VOCABULARY, TERMS, AND PLACES

Use the clues provided to fill in the crossword puzzle below.

Across

2. Hot springs and geysers in Iceland produce ____________ energy.

5. Much of Northern Europe has a mild climate due to the North Atlantic ____________.

6. the capital of the United Kingdom

7. Scandinavian warriors who sailed longboats

8. Britain's legislative body

Down

1. capital of a Scandinavian country known for being neutral for the past 200 years

2. a spring that shoots hot water and steam into the air

3. deep inlet of the sea carved by melting glaciers

4. leading seaport and industrial center, as well as Norway's capital

COMPREHENSION AND CRITICAL THINKING

Read each sentence and fill in the blank with the word in the word pair
that best completes the sentence.

1. Natural resources found in Northern Europe include ___________________.
(timber and fish/glaciers and cliffs)

2. Iceland and Greenland are often considered part of ___________________.
(Scandinavia/the British Isles)

3. British ___________________ has influenced much of the world.
(Parliament/culture)

4. There is conflict in Northern Ireland due to ___________________
differences. **(religious/cultural)**

5. Today, Scandinavia is one of the most ___________________ regions in
Europe. **(war-ridden/peaceful)**

REVIEWING THEMES

Using the lists below, determine what theme from geography they have
in common.

Themes

location	place	region	movement	human-environment interaction

_________________________ **1.** longboats, timber, hydroelectric energy

_________________________ **2.** fjords, long coastline, major seaports

_________________________ **3.** marine west coast, North Atlantic Drift,
wealthy and powerful

REVIEW ACTIVITY: TRAVEL BROCHURE

Create a marketing brochure for a cruise that travels throughout
Northern Europe. Using the political map at the beginning of your
textbook chapter, describe the route and stops your cruise ship would
make. Include photos from newspapers, magazines, or other sources
your teacher approves. Make sure you include information about
climate, physical features, economics, history, and culture.

Vocabulary Builder

SECTION 1

1. glaciers
2. Scandinavia
3. fjord
4. geothermal energy
5. North Atlantic Drift
6. b
7. d
8. d
9. d

SECTION 2

1. Industrial Revolution
2. agreement
3. Stonehenge
4. constitutional monarchy
5. disarm
6. capital of UK, center of world trade, largest city in British Isles
7. Ireland's capital, computer and software industries, located on the coast
8. Great Charter, limited king's power, ordered everyone to obey the law

SECTION 3

Possible responses:

1. The *Vikings* used quick and powerful longboats to attack and conquer villages.
2. Eighty percent of Greenland is covered by ice, so much of the land is *uninhabitable*.
3. Sweden has been a *neutral* country for almost 200 years.
4. They are heated with geothermal energy from *hot springs* and *geysers*.
5. *Stockholm*, Sweden, is built on 14 islands and part of the mainland.

Biography

MARY, QUEEN OF SCOTS

WHAT DID YOU LEARN?

1. possible answer—Her choices made her unpopular with the people of Scotland and led to her being forced to give up the throne.
2. possible answer—With Queen Elizabeth out of the picture, Mary hoped to become Queen of England.

ACTIVITY

Students should create a detailed family tree showing Mary Stuart's relations to other rulers.

Biography

HANS CHRISTIAN ANDERSEN

WHAT DID YOU LEARN?

1. possible answer—He met Jonas Collin, who helped him get the money to continue his education.
2. possible answer—He was born into poverty and had a difficult childhood, but through determination and perseverance he became successful and famous.

ACTIVITY

Answers will vary, but students' interviews should relate details from Andersen's life to one of his fairy tales.

Literature

CALL-OUT BOXES

1. Montague; Capulet
2. She means that a rose would still have the same qualities if it were called something else, just as Romeo would.
3. If Juliet agrees to love him, he will use a new name.

ANALYZING LITERATURE

1. They are in love, but their families are enemies.
2. Students may answer that they will have to hide their relationship or that they will end their relationship out of fear for Romeo's life.

Activity

Students should show an understanding of both iambic pentameter and the sonnet form in their poems.

Primary Source

WHAT DID YOU LEARN?

1. a fjord
2. possible answers—scared, shocked, anxious, tired, sad, aliented, lonely
3. possible answer—very thin, unrealistic
4. possible answer—Nature is distorted to represent the main figure's emotions.

Geography and History

MAP ACTIVITY

1. Students should color England, Scotland, Wales, and Northern Ireland yellow.
2. Students should color Sweden, Denmark, Norway, Finland, and Iceland green.
3. Students should outline at least three fjords in red; most are located on Norway's west coast.
4. Students should place an "X" on the British Isles, northern France, Finland, Iceland, Norway, Germany, Russia, and Sweden.

ANALYZING MAPS

1. about 1200 miles
2. It is part of Ireland geographically, but part of the United Kingdom politically.
3. Its longitude extends from about 10° E to about 25° W. Its latitude extends from about 50° N to about 75° N.
4. Glaciers in the Arctic Ocean account for the jagged west coast. The east coast is enclosed by the Baltic Sea and Gulf of Bothnia, so glaciers had less access, making it a straighter coastline.

Social Studies Skills

PRACTICE THE SKILL

Students' recall (in any order or form) is more important than the writing structure. However, notes should show evidence of freestyle writing for the full 15 minutes.

Geography for Life

1. Denmark, Norway, Sweden
2. They went west from the Faeroe Islands.
3. They went south from Denmark and Sweden.
4. They went to Eastern Europe, Russia, and Asia from Sweden.
5. The latest date on the map is 1190 in the Norman Kingdom in what is now Italy.

Students' essays should demonstrate an understanding of the major issues involved in migration discussed in the activity.

Critical Thinking

1. Answers will vary but should focus on the fact that the Chunnel physically connects Britain to the continent of Europe, effectively ending its isolation.
2. Answers will vary but should show recognition of the headline's suggestion of total isolation from the European mainland.
3. Answers will vary, but students should recognize that the marl was softer than the chalk, which made tunneling easier, but harder than clay or sand, so it provides some rigidity and support for the tunnel.
4. The drawing shows several fault lines. A major shift along any of these faults could warp or snap the tunnel and thus shut it down for a long period, or even permanently.

Focus on Reading

PRACTICING USING CONTEXT CLUES

1. long, winding channels
2. energy produced from the heat of Earth's interior
3. an ancient monument
4. heritage
5. the capital of Norway
6. not able to support human settlement

CREATING YOUR OWN CONTEXT CLUES

Sentences will vary but should show an understanding of the use of context clues.

Focus on Writing

Letters should include information about the geography, history, and culture of Northern Europe, as well as students' reactions to their visit to the region.

Chapter Review

REVIEWING VOCABULARY, TERMS, AND PLACES

Across:

2. geothermal
5. Drift
6. London
7. Vikings
8. Parliament

Down:

1. Stockholm
2. geyser
3. fjord
4. Oslo

COMPREHENSION AND CRITICAL THINKING

1. timber and fish
2. Scandinavia
3. culture
4. religious
5. peaceful

REVIEWING THEMES

1. human-environment interaction
2. place
3. region

REVIEW ACTIVITY: TRAVEL BROCHURE

Brochures should cover the information about Northern Europe in an organized, logical way. Pictures should highlight key stops during the cruise. Brochures should be attractive and persuasive.